FOREST CREATURES

FOXES IN THE FOREST

JENNIFER LOMBARDO

Gareth Stevens PUBLISHING

I0811629

Please visit our website, www.garethstevens.com. For a free color catalog of all our high-quality books, call toll free 1-800-542-2595 or fax 1-877-542-2596.

Library of Congress Cataloging-in-Publication Data
Names: Lombardo, Jennifer, author.
Title: Foxes in the forest / by Jennifer Lombardo.
Description: New York, NY : Gareth Stevens Publishing, 2023. | Series: Forest creatures | Includes index.
Identifiers: LCCN 2022003866 | ISBN 9781538279137 (library binding) | ISBN 9781538279113 (paperback) | ISBN 9781538279120 (set) | ISBN 9781538279144 (ebook)
Subjects: LCSH: Foxes–Juvenile literature. | Forest animals–Juvenile literature.
Classification: LCC QL737.C22 L639 2023 | DDC 599.775–dc23/eng/20220203
LC record available at https://lccn.loc.gov/2022003866

Published in 2023 by
Gareth Stevens Publishing
29 East 21st Street
New York, NY 10010

Copyright © 2023 Gareth Stevens Publishing

Editor: Jennifer Lombardo
Designer: Rachel Rising

Portions of this work were originally authored by David Lee and published as *Creatures of the Forest: Foxes.* All new material in this edition was authored by Jennifer Lombardo.

Photo credits: cover Jim Cumming/Shutterstock.com; cover, p. 1 furtseff/Shutterstock.com; cover, pp. 1, 3–5, 6, 8, 10, 12, 14, 16, 18, 20, 22–24 lavendertime/Shutterstock.com; cover, pp. 1, 3, 4, 6, 8, 10, 12, 14, 16, 20, 22–24 Zvezdesign/Shutterstock.com; p. 5 udaix/Shutterstock.com; p. 7 Josef_Svoboda/Shutterstock.com; p. 9 WildMedia/Shutterstock.com; p. 11 (top) Narit Jindajamorn/Shutterstock.com; p. 11 (bottom) Agarianna76/Shutterstock.com; p. 13 WildlifeWorld/Shutterstock.com; p. 15 Scalia Media/Shutterstock.com; p. 17 Sergey Bezgodov/Shutterstock.com; p. 18 Eric Isselee/Shutterstock.com; p. 19 christian kober/Shutterstock.com; p. 21 JoannaPerchaluk/Shutterstock.com.

All rights reserved. No part of this book may be reproduced in any form without permission in writing from the publisher, except by a reviewer.

Printed in the United States of America

Some of the images in this book illustrate individuals who are models. The depictions do not imply actual situations or events.

CPSIA compliance information: Batch #CSGS23: For further information, contact Gareth Stevens, New York, New York, at 1-800-542-2595.

Find us on

CONTENTS

Words in the glossary appear in **bold** type the first time they are used in the text.

A CLOSER LOOK AT FOXES

There are many foxes in forests around the world. In fact, there are 37 species, or kinds, of fox! However, only 12 of them are "true" foxes. These 12 **evolved** from a common **ancestor** thousands of years ago. The other species are still foxes, but they split off from the true foxes at a later point in history.

Foxes are part of the Canidae family, which includes dogs and wolves. Many species of fox are more similar to wolves or coyotes than they are to the true foxes.

FIND OUT MORE!

Scientists generally give the name "fox" to any small canine that has triangular ears, a flattened skull, a bushy tail, and a pointed snout, or nose.

Scientists classify, or organize, livings things into groups. Animals' scientific names use their genus and species classification. This shows the scientific classification of red foxes.

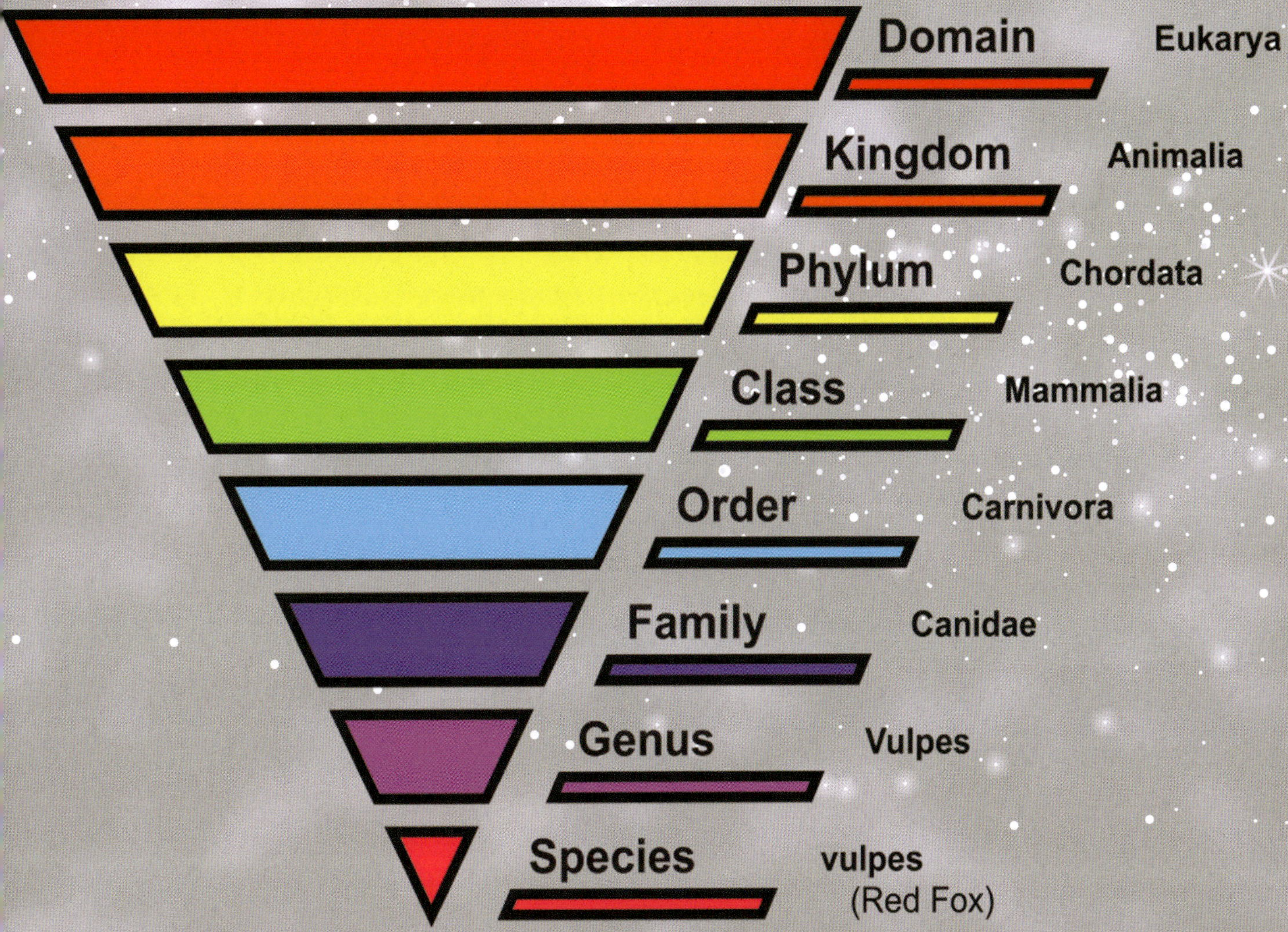

ALL AROUND THE WORLD

The 12 species of true fox are the red fox, Arctic fox, fennec fox, Bengal fox, pale fox, Blanford's fox, Cape fox, corsac fox, Rüppell's fox, Tibetan sand fox, kit fox, and swift fox.

The red fox is the largest species and the most widespread. Its **habitat** includes North America, Asia, Africa, Europe, and even the Arctic. Its favorite foods are mice and rabbits, but it will also eat birds and fruit. It will even steal garbage!

FIND OUT MORE!

Red foxes sometimes have bits of gray in their fur, so they can be confused with gray foxes. However, gray foxes have black-tipped tails, and red foxes have white-tipped tails.

The corsac fox, shown here, lives in Asia.

MAKING A HOME

Not all foxes live in the forest. Arctic foxes live in the Arctic tundra, which is a very cold land with no trees. The kit fox lives in the deserts of the southwestern United States and Mexico. The fennec fox lives in a desert called the Sahara in Africa.

Wherever they live, foxes make homes called dens. Dens give them a place to store food, and they keep the mothers safe while they give birth. Sometimes the foxes sleep in their dens, but not always.

FIND OUT MORE!

A red fox den can be as deep as 8 feet (2 m) and as long as 75 feet (23 m) underground! They have more than one room and often more than one entrance.

Foxes take very good care of their cubs.

NAMES FOR FOXES

Every animal has names that help people **identify** them. A female fox is called a vixen, and a male fox is called a dog or a tod. Baby foxes can be called pups, kits, or cubs.

Foxes live alone except when they're **mating** or raising their young. A group of foxes is called a leash or a pack. They also hunt alone, but sometimes a parent will take their older cubs on a hunt to teach them what to do.

FIND OUT MORE!

Red fox cubs are gray or brown when they're born. They are also blind and deaf for a couple of weeks. They need their parents to survive.

Unlike most foxes, fennec foxes live in family groups. They use their dens to keep them cool in the desert heat.

HOW FOXES ACT

Many different **cultures** have stories about foxes being smart and playing tricks on other animals. One reason for this might be because they're good at **adapting** to their surroundings and at getting away from hunters.

Foxes are also smart enough to communicate with one another. When two foxes meet, they move their ears, tails, and other body parts to "talk" to each other. Foxes also mark trees or rocks with their waste to let other foxes know which land is theirs.

FIND OUT MORE!

Foxes love to play! They'll play with other foxes or other animals, such as deer. They'll even play by themselves. Some golf courses report foxes stealing balls to play with!

Foxes sometimes play by pretending to fight with each other.

HOW FOXES SOUND

Foxes make many different noises. Gray foxes bark like dogs to warn other animals away. Red fox vixens attract mates by making a few short, high-pitched sounds. Cubs bark in a different way to call to their parents.

One of the scariest noises red foxes make is a long scream. People who've heard it say it sounds like a human woman being hurt. Vixens and tods both make this noise when they're ready to mate.

FIND OUT MORE!

In 2013, a band from Norway made a funny song called "What Does the Fox Say?" After that, fox noises became one of the most searched-for animal noises on the internet!

Red foxes can make at least 20 different sounds.

WILD AND TAME

Most foxes are wild animals. They don't hunt people, but they can hurt a person badly if they feel **threatened** or if they have **rabies**. If one comes up to you, it's probably used to people feeding it. However, it'll usually run away if you make a loud noise.

Some foxes are friendlier than others. These are called tame foxes. They're still wild animals, though. A small number of silver foxes are **domesticated**, and these can be kept as pets.

FIND OUT MORE!

In 1959, a Russian man started mating the tamest silver foxes he could find. These foxes passed on that tameness to their cubs. Within 50 years, the tame foxes became domesticated.

Domesticated foxes often enjoy cuddles.

THREATENED FOXES

Foxes live for only three or four years in the wild. Some starve to death because other animals eat all the food before the foxes can. Others, especially the red and Arctic foxes, are hunted for their fur.

One of the biggest threats to all forest animals, including foxes, is deforestation. Deforestation is when people cut down too many trees in a forest habitat. This forces all the forest animals to leave their homes.

Deforestation leaves fewer places for foxes to hide and to find food.

A SHRINKING HABITAT

Another threat to foxes is **climate change**. As the planet warms up, some parts of the Arctic are starting to look more like forest than tundra. This is shrinking the Arctic fox's habitat and killing some of the animals it feeds on.

Red foxes are moving north into these places. They're bigger than Arctic foxes, so they win in fights for food. Some red foxes will even kill and eat Arctic foxes.

FIND OUT MORE!

Even though foxes face threats, most of them aren't **endangered**. In fact, red foxes and Arctic foxes have some of the healthiest populations, or numbers, of all animals on the planet.

Arctic foxes' homes are threatened by climate change.

GLOSSARY

adapt: To change to suit conditions.

ancestor: An animal that lived before others in its family tree.

climate change: Long-term change in Earth's climate, caused mainly by human activities such as burning oil and natural gas.

culture: The beliefs and ways of life of a group of people.

domesticated: Bred and raised for use by people.

endangered: In danger of dying out.

evolve: To grow and change over time.

habitat: The natural place where an animal or plant lives.

identify: To find out the name or features of something.

mate: To come together to make babies.

rabies: A deadly disease that affects the central nervous system. It's carried in the spit of some animals.

threatened: Likely to suffer harm.

FOR MORE INFORMATION

BOOKS

Gardner, Kate, and Ossi Saarinen. *How to Find a Fox.* Philadelphia, PA: RP Kids, 2021.

Murray, Julie. *Fennec Fox.* Minneapolis, MN: Abdo Kids, 2020.

WEBSITES

Animal Fact Guide: Arctic Fox
animalfactguide.com/animal-facts/arctic-fox
Read more about the Arctic fox and look at some cute pictures!

Easy Science for Kids: Fox—Clever and Shrewd
easyscienceforkids.com/all-about-foxes
Learn more about the red fox.

Publisher's note to educators and parents: Our editors have carefully reviewed these websites to ensure that they are suitable for students. Many websites change frequently, however, and we cannot guarantee that a site's future contents will continue to meet our high standards of quality and educational value. Be advised that students should be closely supervised whenever they access the internet.

INDEX